Nazareth Koko

Diversity and Management of Non-Timber Forest Products (NTFPs)

Nazareth Koko

Diversity and Management of Non-Timber Forest Products (NTFPs)

in the commune of PISSA in the Central African Republic

ScienciaScripts

Imprint

Any brand names and product names mentioned in this book are subject to trademark, brand or patent protection and are trademarks or registered trademarks of their respective holders. The use of brand names, product names, common names, trade names, product descriptions etc. even without a particular marking in this work is in no way to be construed to mean that such names may be regarded as unrestricted in respect of trademark and brand protection legislation and could thus be used by anyone.

Cover image: www.ingimage.com

This book is a translation from the original published under ISBN 978-620-2-54530-3.

Publisher:
Sciencia Scripts
is a trademark of
International Book Market Service Ltd., member of OmniScriptum Publishing Group
17 Meldrum Street, Beau Bassin 71504, Mauritius
Printed at: see last page
ISBN: 978-620-3-17341-3

I dedicate this dissertation to my parents *Marcel KOKO* and *Rosalie NDINGFEI*.

THANKS

This work cannot pretend to be the fruit of our exclusive will and action. Indeed, several people have contributed to it from near or far. May all of them find here the expression of our sincere thanks.

I would like to thank very much Dr **Marie Louise AVANA-TIENTCHEU,** who despite her heavy academic and administrative workload has kindly accepted to direct the work of this research. May she find here the expression of our thanks.

I would also like to present my gratitudes to Dr **Marie-Caroline MOMO SOLEFACK**, for her contribution to the success of this research work.

We also take this opportunity to thank all the teachers of the Department of Plant Biology of the University of Dschang more particularly, Dr. **NGUETSOP F,** Head of the said Department; Pr. **FONKOU. T** head of the Laboratory of Applied Botany; Drs **KAMENGNE. G,** and **PINTA J**.

We would also like to thank you very warmly :

- **The NGO AGIR-CENTRAFRIQUE**, which provided me with a whole team of its executives who helped us to collect data in the field in the forest area of the Commune of PISSA ;
- Dr. **Denis BEINA** and **Bruno de SEMBOLI**, who provided us with specific and rich documentation related to the object of our research. May they find here all our gratitude.
- The **KOKO** family, my son **Jospin,** not forgetting my fiancé Yvon Dimitri BRIGUETTE, who never ceased to support me morally during all the periods of this training.
- Finally, to all those who could not be quoted in these few paragraphs, may they find here the expression of our deep gratitude.

TABLE OF CONTENTS

THANKS — *2*

TABLE OF CONTENTS — *3*

LIST OF ABBREVIATIONS — *6*

SUMMARY — *8*

ABSTRACT — *9*

CHAPTER I. INTRODUCTION AND REVIEW OF THE LITERATURE — *10*

INTRODUCTION — *10*

I.1.   Background — *10*

I.2.   Research problem — *11*

I.3.   Main and specific research objectives — *11*

LITERATURE REVIEW — *13*

I.4.   DEFINITION OF CONCEPTS AND TERMS — *13*

II.1.1.   Biodiversity — *13*

II.1.2.   Non-Timber Forest Products (NTFP) — *13*

II.1.3.   NTFPs of plant origin — *15*

II.1.4.   Management — *15*

I.5.   CATEGORIES OF PFNL OF PLANT ORIGIN — *15*

II.2.1.   Food NTFPs — *16*

II.2.2.   Medicinal NTFPs — *16*

II.2.3.   NTFPs used in other services — *17*

I.6.   Importance of non-timber forest products — *17*

I.7.   NTFPs and Poverty — *18*

I.8.   Riparian population and knowledge of NTFPs — *19*

I.9.   NTFP management strategies and conservation — *20*

CHAPTER III MATERIALS AND METHODS III.1.GEOGRAPHICAL FRAMEWORK OF THE STUDY AREA — *21*

III.2.   PHYSICAL ENVIRONMENTS — *22*

III.2.1.   Relief — *22*

III.2.2.   Climate — *22*

III.2.3.	Vegetation	22
III.2.4.	Soil	22
III.2.5.	Hydrography	23
III.3.	HUMAN AND SOCIO-ECONOMIC ENVIRONMENT	23
III.3.1.	Population and socio-demographic aspects	23
III.3.2.	communication channels	23
III.3.3.	Socio-economic activities	24
III.4.	SURVEY METHODS AND TOOLS III.4.1.METHODS USED	24
III.4.1.1.	Documentary research	24
III.4.1.2.	Choice of villages.	25
III.4.1.3.	Choice of respondents	26
III.4.2.SURVEY TOOLS USED		26
III.4.2	Interviews	26
III.4.2	2.2 Questionnaire	26
III.4.2 3.Field Observation and Sample Collection		27
III.4.2	4 Identification of samples	27
III.4.3	DATA ANALYSIS METHOD	28
III.4.3 1.Evaluation of the most useful species		28
III.4.3 2.Quantification of ethnobotanical data		28
a. Use value of a species		28
a.	Diversity Index	28
CHAPTER IV RESULTS		29
IV GENERAL CHARACTERISTICS OF RESPONDENTS		29
IV.1.	FLORISTIC DIVERSITY OF THE NFPS IN THE MUNICIPALITY OF PISSA	29
IV.2 1.Families represented in the municipality of PISSA		29
IV.2.2.	Genders cited in the municipality of PISSA	30
IV.2.3.	Species cited in the municipality of PISSA	31
IV.2 4.Representativeness of households surveyed on NTFPs		32
CHAPTER V DISCUSSION AND CONCLUSION		40
V.1.DISCUSSION		40

LIST OF ABBREVIATIONS

ASCNA   Agency for the Safety of Air Navigation in Africa
CDB   Convention on Biological Diversity
COMIFAC Central African Forest Commission
FAO   Food and Agriculture Organization of the United Nations
IGN.............. National Geographic Institute
MEFCPE   Ministry of Water, Forests, Hunting and Fishing, in charge of the
Environment
WHO ....World Health Organization **NGO**
Non-Governmental Organization **LD**   **P**
Local Development Plan **NTF**   **P** Non-
Timber Forest Product **CAR** Central African
Republic
EWC.............. Democratic Republic of Congo
IUCN World Conservation Union

SUMMARY

This study was carried out on the "Diversity and Management of Non-Timber Forest Products (NTFPs) in the commune of PISSA in the Central African Republic" from April to June 2014. The work consists of contributing to the sustainable management of NTFPs in CAR in general and in the commune of PISSA in particular.

The methodology used consists of direct semi-structured surveys using a pre-established questionnaire, allowing 79 people to be interviewed, 61% of whom are men and 39% women, divided into four ethnic groups: the Ngbaka (59%), the Mbati (20%), the Aka pygmies (12%), and the Bofi (9%). The majority of respondents are literate.

The results of this survey reveal 102 plant species, divided into 78 genera and 42 families used in the municipality of PISSA, of which 50% are used in food, 24% used in traditional pharmacopoeia and 26% used in other services. The most used plant organs are leaves, bark, fruit, seeds and stems; the most common types of use are snacks, vegetables, condiments, tea and packaging. However, NTFPs are harvested in the forests more than 5 km from the villages surveyed through collection, gathering and felling; the plants are rare at the place of harvest for the medicinal and service categories, while they are abundant for the food category. Harvested NTFPs regenerate naturally while seed regeneration and vegetative propagation are less cited by the surveyed populations. It appears from this study that the surveyed populations do not know of any method to conserve the resources they harvest and very few use their traditional method for the conservation of these resources.

Keywords: Plant NTFP, Management, Diversity.

ABSTRACT

This survey has been made on" the Diversity and Management of the non Woody Forest Products (PFNL) in the township of Pissa in Central african Republic " of the month of April to the June 2014.Le work consists to contribute to the lasting management of the PFNL in general in RCA and in the township of PISSA in particular. The used methodology consists in making direct investigations semi structured with the help of a questionnaire pre- establishes, permitting to interrogate 79 people of which 61% men and 39% women, distributed in four ethnic groups that are the Ngbakas (59%), the Mbatis (20%), the Pygmy Akas (12%), and the Bofis (9%). They investigated to majority know to read and to write.

The results of this investigation reveal 102 plant species, left in 78 kinds and 42 families used in the township of PISSA, among which 50% are used in food, 24% used in the traditional pharmacopeia and 26% used in the other services. The plant organs the more used, are the leaves, the peels, the fruits, the seeds and the stems; the types of most quoted uses are amuse them muzzles, the vegetables, the condiments, tea and the packings. However the harvests of the PFNL get used in the drills to more of 5 km of the villages investigated by the withdrawal, the pickup and slaughtering; the plants are rare instead of harvest for the categories to medicinal uses and services, whereas they are abundant for the food category. The PFNL harvested regenerate naturally while regeneration by the seed and the vegetative multiplication is less quoted by the populations investigated. He/it is evident from this survey that the populations investigated know no method to keep resources that they exploit very little use their traditional method for the conservation of these resources

CHAPTER I. INTRODUCTION AND REVIEW OF THE LITERATURE

INTRODUCTION

I.1. Background

Tropical rainforests are the most complex and diverse ecosystems on the planet (Puig, 2001) and are particularly important in terms of species richness as well as concentration of endemic species or species not found elsewhere on Earth (Brooks & al., 2006).

This counterpart, the forests of the Congo Basin, constitute the second largest massif of tropical forests after the Amazonian massif. With an estimated total area of about 200 million hectares, or nearly 91% of Africa's dense rainforests, they represent the main forest resources of the entire continent. (CBD & COMIFAC, 2010).

According to Gueneau, 2005 65 million people live in or near these forests, which are home to a heritage of animal and plant species of great diversity. This plant diversity is largely made up of non-timber forest products (NTFPs). According to the FAO in 2007, a non-timber forest product can be defined as any good of biological origin other than wood, derived from forests, other wooded land, and trees outside forests. They are also sources of food, shelter and income for more than 30 million people (Bahuchet, 1995).

However, the economic stagnation and poverty prevailing in most tropical countries have led to ever-increasing pressure on forest resources. This pressure is leading to intense deforestation which deprives local populations of important sources of subsistence and income. It also affects climatic conditions and threatens the existence of thousands of animal and plant species (Kahn, 1988; Varquez & Gentry, 1989).

In the tropics in general, and in CAR in particular, many rural people depend on non-timber forest products (NTFPs) for their livelihoods and source of income. However, the availability of these NTFPs is declining due to land clearing to increase arable land (Burnley, 1999).

According to Clark & Tchamat in 1998, many species are increasingly threatened due to habitat loss, destructive and excessive harvesting methods that contribute to reducing wild populations below the regeneration threshold, and insufficient domestication of NTFPs.

I.2. Research problem

The Central African Republic is still one of the countries where the burden of poverty is enormous. The population lives mainly from the exploitation of natural resources. This perception is becoming worrying, especially with the population growth and the scarcity of natural resources. However, the ignorance of the potential of some NTFPs due to the lack of inventory, and the limited capacities of actors to access information on NTFP markets at the local and regional level, contribute to the underdevelopment of the NTFP sector. In this regard, Konzi & al. point out that, the lack of scientific or in-depth knowledge on the majority of these products, particularly on the availability of the resource, on appropriate harvesting methods, as well as on conservation methods hinders this sector, (KONZI & al., *2010*).

The socio-economic importance of NTFPs in general is not well integrated into the process of sustainable management of forest resources. In spite of their importance and their enormous potentialities, in CAR, we note a very low valuation, a low knowledge of the resource, a lack of information on the role of NTFPs in the household economy. To achieve this, this investigation attempts to answer the following questions:

✓ What are the main NTFPs in the locality and their traditional uses by the population?

✓ What are the local methods of NTFP management (exploitation, regeneration methods, conservation methods, etc.)?

✓ What are the main niche markets?

✓ What is the farmer's perception of the state of resources providing NTFPs?

I.3. Main and specific research objectives

We this study consists of Contributing to the sustainable management of NTFPs in CAR in general and in the commune of PISSA in particular, it is about :

✓ To inventory and categorize NTFPs of plant origin that are a priority for the locality

✓ Identify local management methods for exploited NTFPs (exploitation strategies, harvesting methods, regeneration methods, conservation methods, etc.).

✓ To identify the plant organs and types of use most common in this locality ;

✓ Identify NTFP collection niches (Forest, behind the boxes, fields);

✓ Assess peasant perceptions (know-how) on the state of resources providing NTFPs (Abundant, less abundant, rare).

This work is subdivided into four chapters plus the portion of the Appendix. Chapter one focuses on the generality of NTFPs. Chapter two, on the other hand, focuses on the methodological framework of the research; chapter three is devoted to the presentation of results. Finally, Chapter Four, however, deals with discussion and conclusion.

LITERATURE REVIEW

I.4. DEFINITION OF CONCEPTS AND TERMS

II.1.1. Biodiversity

In just a few years, the term "Biodiversity" has become an unavoidable term in scientific circles and among the general public. The question of biological diversity is not a new idea, however, because the natural sciences have long been interested in the inventory, origin, and dynamics of the living world (Lévêque, 1998).

In Central Africa, the Congo Basin, which represents 3/10th of the water resources of the African continent, is known for its diversity (Sanwa, 2010).

According to the CBD in 2008, biodiversity, or biological diversity, is understood to be "the diversity of life on the planet, including genetic diversity, the diversity of species and the diversity of ecosystems. It combines three elements, namely the conservation of biodiversity, the sustainable use of natural resources and the equitable sharing of the benefits arising therefrom".

In CAR, the environmental code defines it as follows: "Biological Biodiversity is the variability of living organisms from all sources including, among others, terrestrial, marine and other aquatic ecosystems and the ecological complexes of which they are part (**MEE, 2OO7**).

II.1.2. Non-Timber Forest Products (NTFP)

A non-timber forest product can be defined as: "Any good of biological origin other than wood, derived from forests, other wooded land, and trees outside forests". (FAO 2007). For Belcher, 2009, non-timber forest products (NTFPs) are usually defined as: "forest products of plant or animal origin other than timber and fuelwood. They are also defined as goods and services, other than timber, derived from renewable forest resources and which enable people to meet their basic needs (food, health, construction, handicrafts, socio-cultural) and whose marketing benefits primarily village communities".

On the other hand, COMIFAC defined it in 2008 as: "Spontaneous forest products of plant origin whose local importance has been demonstrated, as well as certain species of microfauna such as caterpillars whose life depends essentially on commercially important forest species (Examples: Ayous (*Triplochyton scleroxylon*) and Sapelli

(*Entandrophragma cylindricum*)" However, donors refer to them as NTFPs, which means "non-timber forest products" in full. Given their economic importance in relation to that of logs, they are often referred to as "forest products".

The terms "minor" or "secondary" are used.

For PODA in 2012, all NTFPs represent products with High Conservation Values, as they contribute to the livelihoods of people in African countries and bring them substantial revenues. Fungi are considered an independent kingdom of vascular plants, and therefore should be included as a third category in the classification of NTFPs (Toirambe, 2005). In Congo, this activity is not very developed. It is mainly products used for traditional handicrafts and products used for construction that are concerned (Loubelo, 2012).

II.1.3. NTFPs of plant origin

NTFPs of plant origin are sources of food and are consumed as a staple or main course, side dish, binder, condiment or as aromatics, stimulants or aphrodisiacs, "appetizers". In higher plants, these are various plant organs including: buds, leaves, stems, bark, roots, bulbs, underground rhizomes and tubers, fruits and seeds. In lower plants, fungi are consumed mainly (FAO, 2004b).

II.1.4. Management

Sustainable forest management is a vast subject that comprises two aspects: the legal tools enabling the implementation of a sustainable forest management policy, and the documents enabling the logger to organise this same sustainable management within a concession. There are generally two complementary approaches to the sustainable management of NTFPs: ensuring the sustainable exploitation of these resources in their natural environment (*In situ*) and encouraging cultivation in agricultural areas (*Ex situ*) (CBD, 2008). Sustainable management of forests ensures their biological diversity, productivity, regeneration capacity, vitality and their ability to satisfy, now and in the future, relevant economic, ecological and social functions, at local, national and international levels, without causing damage to other ecosystems (CRPF, 2012).

I.5. CATEGORIES OF PFNL OF PLANT ORIGIN

NTFPs include a wide variety of useful products that can be divided into two main categories: goods and services.

According to the Global Forest Resources Assessment, FAO classifies plant NTFPs into eight categories, namely food, fodder, raw material for the preparation of medicines and aromatic products, raw material for the preparation of dyes and colorants, raw material for utensils, handicrafts and construction, ornamental plants and other plant products (FAO, 2004b).

II.2.1. Food NTFPs

According to FAO, 2003, food NTFPs are non-timber products of biological origin derived from forests, other wooded land and trees outside forests and intended for food, feed, food processing and marketing. They can be harvested from the wild, or produced in forest plantations, agroforestry areas, or by trees outside forests.

Another key point to note here is the distinction that is made between a food product called NTFP, because it comes from trees of spontaneous origin, and that from domesticated trees, identified as an agroforestry product (Simons & Leakey, 2004). In higher plants, the various plant organs most often exploited for food are: fruits, nuts, bark, leaves, seeds, roots, tubers, beans and oils. In the lower plants, edible fungi are the most exploited species. They appear in nature only seasonally and are a prestigious consumer product that is highly appreciated by the local population. (FAO, 2004 b). In other words, non-timber forest products are not only an essential source of food, but also the basis for the economic activities of millions of families (CIFOR, 2012).

II.2.2. Medicinal NTFPs

Medical NTFPs are much more used in the sanitary but also food framework, the World Health Organization (WHO) has pointed out. According to the WHO, 2012, more than 80% of African populations use traditional medicine and pharmacopoeia to address health problems (Sawadogo, 2010; Kolling, 2010; Mangambu, 2013). Despite current advances in biology and medicine, the majority of people in developing countries do not have access to adequate health care due to weak economic systems (Konda & al., 2011; Singh & Singh,2012; WHO, 2012). The use of readily available local resources would constitute a real palliative solution in the perspective of the Millennium Development Goals (Kumar & Lalramnghinglova, 2011; Mangambu & al., *2012).*

Non-timber plant products potentially contribute to the supply of raw materials from which active ingredients are extracted by the pharmaceutical industries (25% of prescriptions in the United States recommend drugs containing plant extracts). The African continent abounds in a wide variety of medicinal plants (Mangambu & al., 2010). 70% of local communities use a variety of forest products including

NTFPs for their health care, In CAR, these products are widely used because very few modern health centers exist in rural areas (FAO, 2006a).

II.2.3. NTFPs used in other services

The use of NTFPs is diversified, i.e. multi varied. For example, NTFPs are mainly used for construction, basketry, clothing, hunting, fishing, etc.. The most used parts in this case are wood, followed by leaves, small diameter stems, bark, gums, resins and fruits.

Wood is used as posts and stakes in the framework for constructions, leaves and lianas are also used in construction (roofing, binder in the framework, etc.). The leaves of some palms (*Elaeis, raffia, Sclerosperma, .*) and herbaceous (*Aframomum giganteum),* are used to cover roofs: the case of vegetable tiles obtained by braiding the long leaflets of leaves. (FAO, 2006a). The leaves of species of the Maranthaceae family (*Ataenidia, Maranta, Megaphrynium, Hupselodelphis*) are often used to wrap cassava bread.

In the handicrafts we find tams-tams (with wood from *Fagara hetzii, Ricinodendron heudelotii,*), canoes (with wood from *Dacryodes spp, Entandrophragma cylindricum, Piptaniastrum africana,* etc ...). The stems of small diameters or rotors come almost from the two genera of vine palms that are: *Eremospatha* and *Laccosperma of the* family Arecaceae. The larger rotors are used as rigid axes for furniture or baskets (Tapsoba, 2011).

I.6. Importance of non-timber forest products

Forests provide goods and services that are essential for 1.2 billion people worldwide (FAO, 2004a). The importance of NTFPs is well established. They are the most obvious manifestation of the value of the forest to local people and are therefore an important factor in the conservation of forest resources.

Socially, more than one million people worldwide rely on forest resources for their livelihood or to supplement their livelihood. In Africa, 60-80% of the needs of the poor depend directly on these natural resources (IUCN, 2003). For these populations, the forest is the main source of energy and food.

NTFPs are among these sources of energy and food for the population. The FAO estimates 65% of the NTFPs used for food in CAR NTFPs often come from the following sources

fortify diets that would otherwise be tasteless and nutritionally deficient (FAO, 2011).

Indeed, from the 1990s to the present day, numerous studies have shown that communities depend on NTFPs for their subsistence and financial needs. The FAO estimates 65% of the NTFPs used in food in CAR NTFPs contribute to the fight against poverty, to the food balance and to the security of populations in urban and rural areas (Mbolo, 2006). In both rural and urban areas, many households live on these products for food on the one hand and for use as a source of income and medicine on the other.

I.7. NTFPs and poverty

NTFPs play a significant role in the fight against poverty, the existence of local communities by providing them with food and income case of mushrooms (Congo), bark of Prunus africanus (Cameroon, Equatorial Guinea), *Rauvolphia vomitora,* and *Pipper guinesse (CAR),* leaves of *Gnetum* sp and Marantaceae (CAR, Gabon, Congo), Rattan (Congo, DRC, Gabon and CAR)

In other words, the forestry sector occupies a prominent place in CAR's poverty reduction strategy (MEPCI, 2007) and contributes in a way significative to the national and local economies of CAR.

Consumption of the majority of NTFPs is on the rise mainly because they are available at low prices. They are becoming an important part of the livelihood strategy of rural communities, particularly given the low rates of economic growth and irregular agricultural production (Bonané, 2006).

It is important to remember that NTFPs are of great importance to the population of Central Africa. In the countries of the Congo Basin, the development index varies between 0.361 and 0.703, 60 to 80% of the needs of poor populations are directly met from these forest resources (COMIFAC, 2010). However, these forests remain among the most poorly known on a global scale and, at the same time, among the most threatened. These threats are not only to the forest itself and the species it contains, but also to indigenous societies and their traditional knowledge. Major threats to biodiversity include fragmentation, habitat loss, invasive species and pollution (Sonwa, 2010).

Whether at the local, national, regional and international level, NTFPs provide food, medicine, energy, building materials, construction equipment, and other goods.

fishing, goods and various utensils to the populations. Also, they have great socio-cultural and religious value in the subregion (FAO, 2006b). As such, NTFPs contribute both to food security and the general welfare of the populations in Central Africa. One of the World Bank studies has shown that about 90% of the poorest populations depend on forests for subsistence and income (Catie & al., 2006) and 80% of people living in developing countries use NTFPs to meet a number of needs such as food, health care and income generation (Sama & al., *2010*).

I.8. Riparian population and knowledge of NTFPs

The Central African Republic (CAR) is located at the northern limit of the Congo Basin forest massif. Due to its geographical position and the diversity of its ecosystems, the country abounds in a fairly rich biodiversity, among which are the species providing NTFPs... The populations living in the CAR in general and in the Commune of PISSA have an in-depth knowledge of the non-timber forest products that surround them and the uses of food and medicinal plants. This know-how constitutes an indispensable prerequisite for the sustainable management of these resources, whether it is a question of preserving the productive species in the face of their exploitation. The forest has always been one of the fundamental components of people's living environment because of the gathering, hunting and wood products they harvest there.

Food sources of NTFPs are consumed as a staple food or main course, supplementary food, binder, condiments, or as aromatics, stimulants, or aphrodisiacs,

"appetizers". In higher plants, these are various plant organs including: buds, leaves, stems, bark, roots, bulbs, rhizomes, underground tubers, fruits and seeds. In the lower plants, fungi are mainly consumed (FAO, 2007).

I.9. NTFP management and conservation strategies

Given the important current and potential contribution of NTFPs to food security and the economy in Central Africa, a sub-regional natural resource management strategy is needed. The main ideas of the current situation on the legislative and regulatory framework date back to the Rio Summit in 1992 when many countries became aware of environmental issues, sustainable management and conservation of forest ecosystems (COMIFAC, 2010) This was followed by other major events with the Millennium Development Goals (MDGs) and the Johannesburg Summit in 2002. In Central Africa, the first initiatives towards regional coordination in the field of biodiversity conservation and sustainable forest management were launched in 1999 with the Yaoundé Declaration, signed by the Heads of State of the six forest countries of the region (De Wasseige & al., 2012).

Riparian and forest populations of CAR and more particularly those of the Commune of PISSA have a thorough knowledge of the non-timber forest products that surround them and the uses of food and medicinal plants. On this subject, Tchatat & Ndoye have written that, "This know-how is an indispensable prerequisite for the sustainable management of these resources, whether it is a question of preserving productive species from commercial exploitation or of improving them. This body of local knowledge and simple conservation techniques can have a positive influence impact on the sustainable management of NTFPs (Tchatat & Ndoye, 2008).

However, the sustainable and responsible exploitation of the resources providing NTFPs combined with their domestication, sustainable harvesting methods and commercialization are avenues to be explored in order to better value them.

CHAPTER III MATERIALS AND

METHODS III.1.GEOGRAPHICAL FRAMEWORK OF THE

STUDY AREA

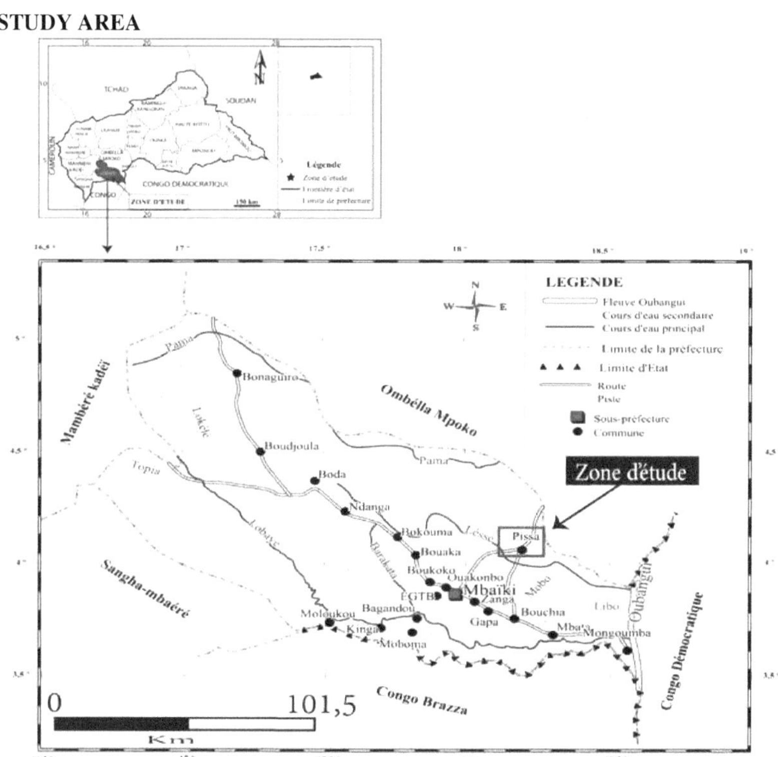

Figure 1: Location of the commune of PISSA in the prefecture of Lobaye

The Central African Republic is a country located in the center of the African continent, covering an area of 623,000km2. Located in the southwest of the CAR in the prefecture of Lobaye, 42 km from the capital Bangui, the commune of PISSA covers an area of 800 km2. It lies between 4°01'12" and 4°05'24" North latitude.

.18°12'36" and 18°15'31" East longitude, bounded to the north by the commune of Lesse, to the south by the commune of Mbata, to the east by the commune of Bimbo and to the west by the commune of Mbaïki (IGN, 2013).

III.2. PHYSICAL MEDIA

III.2.1. Relief

The geological substratum of Pissa dates from the Precambrian. It is generally made up of carbonates from the Upper Proterozoic, this basement is surmounted by blanket formation made of sandstone, dolomite limestone and alluvium from the Quaternary. The result is a flat relief sporadically filled by hills. The Mpere hill to the north-west of Pissa at 430m is the highest point that can be encountered (Bekoye, 2003).

III.2.2. Climate

Situated in the forest massif of south-west Central Africa, the commune of Pissa enjoys a Guinean forest or sub-equatorial climate of the Congo Basin with rainfall ranging from 1500mm to 1800mm per year, spread over 8 to 9 months with a dry season of 2 to 3 months (December-February), the average temperature is 30°C with a thermal amplitude of 2.7°C (PDL, 2006). The rainy season is from March to November while February is considered an off-season month (ASCNA, 2011).

III.2.3. Vegetation

The vegetation of the municipality of Pissa is composed mainly of dense semi-deciduous forest, which is characterized by a high and lush, closed, multi-strate vegetation, rich in liana and epiphyte. The commune of Pissa is located in dense tropical forest in *Triplochiton* and *Terminalia*. There are also deciduous heliophilic species due to the secondary forest. Savannah species such as *Borassus flabellifer* form perceptible plant populations at the northern entrance of Pissa. In these forests many leaves are useful for human food, this is the case of *Gnetum africanum* and *Dorstenia scafegira*. (Boulevert, 1986).

III.2.4. Sol

The pedological formation of the municipality of Pissa includes impoverished ferralitic soils, very represented in some villages, and leached hydromorphic soils. These are characterized by low exchangeable cation capacity and limited fertility, requiring appropriate fertility management in the context of certain agricultural activities (PDL, 2006).

III.2.5. Hydrography

The commune of Pissa also has an important hydrographic network with the Lesse, Mboma, Mbeko, Gbakale, Satombe, Mombou, and Magouga as main rivers. The Mbeko river forms the natural boundary between the communes of Pissa and Mbata (PDL, 2006). The Lesse River passes near Boyali, north of Pissa at 4°6 North and 18°14 East. It flows in the direction of Oubangui in a southeasterly direction. It is a permanent watercourse; Gbatembe and Gbakabe are remarkable swampy arms in the rainy season. During the dry season, their beds dry up in places, the MBeko river is the border in the south of the commune. It flows from west to east and flows into the Lesse (Bekoye, 2003).

III.3. HUMAN AND SOCIO-ECONOMIC ENVIRONMENT

III.3.1. Population and socio-demographic aspects

The Commune of Pissa has 23,792 inhabitants spread over 39 villages. The original population of the area is made up of AKA pygmies. Thus the Ngbakas, who are said to originate from the Congo Basin, practiced semi-nomadism for a long time before settling permanently in the Lobaye region. Today the Commune of Pissa, because of its dynamism and its proximity to the capital Bangui, attracts a large and varied population (PDL, 2006). The population of the Commune of Pissa is made up of ten (10) ethnic groups: the Ngbaka, Banda, Issongo, Gbaya, Mbemou, Zandé, Yakoma, Gbanziri, Kaba and Ali. There is also the presence of some Chadian and Congolese migrants from the DRC. The Ngbakas are in the majority, followed by the Issongo (Mbati), Banda, Gbaya, and others (Mbabale, 2011).

III.3.2. channels of communication

Of the 39 villages in the Commune of Pissa, 35 are located along the national road N°6 on the Bangui-Mbaïki axis. The others are connected by generally well-maintained tracks and are therefore easily accessible. The quality of the road network and the proximity of Bangui mean that transport activities are highly developed in the commune.

III.3.3. Socio-economic activities

The local economy is based on the informal sector. The activities that animate this sector are agriculture, breeding, trade, handicrafts, transport and incidentally, hunting, fishing and processing activities. The formal sector is limited to commercial services, public and social administrative services which are in fact decentralized services and structures of the State with a significant financial impact on the local economy. Commercial activities occupy an important place in the economy of the area. Agriculture and extra-agricultural activities account for 80% of the active population. Despite strong local demand due to the proximity of Bangui and Mbaïki, agriculture remains largely self-consumption with a cultivated area per family of 1 ha. It follows a process of land clearing on burnt land and is characterized by low productivity, the use of rudimentary equipment and the absence of fertilizers and phytosanitary products constitute a major handicap for producers. The lack of mastery of the processes of conservation or processing of agricultural products forces some farmers to sell their products quickly (Mbambale, 2011).

In addition, the use of rudimentary means of transport to evacuate the productions from the fields to the habitat or to the place of marketing also represents an important constraint, the productions are thus generally evacuated from the fields in small quantity which does not allow to envisage a significant rate of marketing. Products from extra-agricultural activities (gathering, fishing, hunting) are sometimes surplus and contribute to the supply of the capital Bangui (PDL, 2006).

III.4. SURVEY METHODS AND TOOLS
III.4.1.METHODS USED
III.4.1.1. Documentary research

Documentary consultations were carried out in the FAO libraries in CAR and on the Internet. We consulted general works, specific books and scientific articles dealing with NTFP issues in general; i.e. their diversity, their economic and social importance in forest regions, their conservation, management, marketing, use and impact of forest exploitation.

III.4.1.2. Choice of villages.

The choice of villages takes into account the level of involvement of the population in the management of NTFPs and also the population density in the villages. Out of 38 villages in the commune of Pissa, only 27 villages grouped into 9 large villages were selected and visited for collection (Table). This makes a 98.75% completion rate, which is very representative.

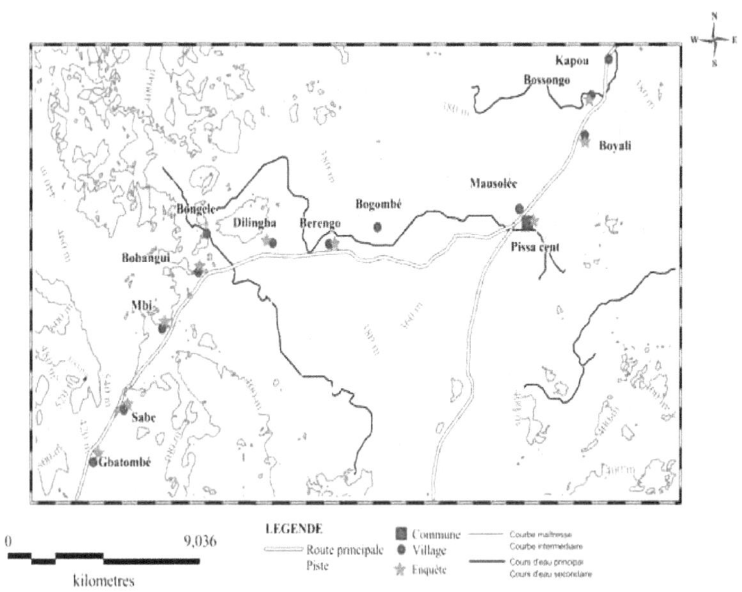

Location of the surveyed villages in the commune of Pissa

Table 1: Sampling of family units in PISSA villages

Villages surveyed	Number of employees Population of the Villages	Number of family units	Number of family unit surveyed	Sampling rate (%)
Pissa Centre	3.844	130	20	15,38
Bossongo	2.020	110	11	10
Boyali	167	45	8	17
Dirigba	302	34	8	23,52
Berengo	455	30	8	26,66
Bobangui	321	47	10	21,27
Mbi	193	27	6	22,22
Sabé	231	26	6	23,07
Gbatombé	336	12	2	16,66
TOTAL	7869	461	79	17,13
	98,75%			

III.4.1.3. Choice of Respondents

This choice was made in a random stratified manner in order to take into account the different levels of informants, the stratification was based on gender and socio-professional categories related to the management of NTFPs.

III.4.2.SURVEY TOOLS USED

III.4.2 1. interviews

We met with the mayor of the commune and the village chiefs to show the purpose of our visit, which made it easier for us to talk with the people in charge of different groups related to the management of NTFPs and the populations of the said commune. In other words, the conversation usually takes the form of a series of questions that are essentially verbal. Thus, the questions that we asked our respondents were about :

- A Brief History of the Municipality of Pissa
- Sustainable management of NTFPs in the Commune of Pissa ;
- The role of NTFPs in the Commune of Pissa ;
- Types of NTFPs for food in the Commune of Pissa ;
- Types of NTFPs for medical and other services in the Commune

III.4.2 2) Questionnaire

In addition to the interviews, we used the pre-established questionnaire to collect information from our respondents.

A total of 79 questionnaires were distributed in the commune surveyed. These surveys were based on semi-structured direct questions about:

- Species providing NTFPs in the locality (in order of importance).
- The different uses of plant NTFPs (food, medicinal and other services)
- The plant organs used (Leaves, Stems, Fruits, Seeds, Roots etc....)
- Management methods, (Conservation, Exploitation, Harvesting etc.)
- Peasant perception of the state of resources (Very abundant, Abundant, Less, Rare).
- The distance to the Sample location
- NTFP collection niches (Forest, Field, Behind the Boxes).

III.4.2 3.Field Observation and Sample Collection

Based on the answers and explanations gathered from the people surveyed, we went to the different plant formations in the study area with a guide for collecting samples of NTFPs. Samples (stems, leaves, roots, tubers, flowers and fruits) as complete as possible were collected on the ground after identification of some of them. The harvest was always accompanied by a herbarium board for the establishment of the control herbarium that follows. The harvested samples are placed between sheets of newspaper as they are harvested and carefully squeezed between the plank presses to dry for about 48 hours.

III.4.2 4 Sample identification

The field survey was accompanied with photo taking of our samples, the NTFPs collected were identified on site in the field for some of them using the practical identification guides. The others were later identified using the different floral species which are :

- The dendrology manual for dense forests in the Central African Republic (Souane, 1989);
- The guide to the practical identification of the main trees in the dense forest of Central Africa (Tailfer, 1989a, 1989b) ;
- The flora of Cameroon (Vivien and Faure., 1985).

III.4.3 DATA ANALYSIS METHOD

At the end of our descent in the study locality, we counted 102 NTFPs used in the said commune. The raw data obtained in the field were manually processed and synthesized in Excel 2007 software sheets for descriptive analyses. Percentage calculations and some indices were made to determine the proportions of products used and the types of uses.

III.4.3 1.Evaluation of the most useful species

NTFPs and their uses have been identified and categorized (food, medicinal, and other services). For each use category, analyses were made to highlight the floristic diversity (number of citations of families, genera, species).

III.4.3 2.Quantification of ethnobotanical data

In order to quantify the ethnobotanical data, calculations of certain indices such as the diversity index (DI) and ethnobotanical use values (OUV) were made to assess the diversity of use, and the level of knowledge of NTFPs in the locality.

a. Use value of a species

It allows to estimate the value placed on the species by the respondents. It is equivalent to the number of uses of the species according to informant i divided by the   total number of informants.

questioned about the species.

$$UV_s = \frac{\sum_{i}^{n} U_{is}}{n_s}$$

With, UVs: use value of a species

Uis :   name of use of the species according to the informant

ns: total number of informants asked about the species.

a. Diversity Index

The diversity index (DI) measures how many respondents use the species and how this knowledge is distributed among the respondents (Codjia *et al.*, 2014).

$$ID = \frac{1}{\sum P_e^2}$$

WithID= Diversity Index

P = The number of uses cited by respondents e

= One species

CHAPTER IV RESULTS

IV GENERAL CHARACTERISTICS OF RESPONDENTS

Our investigations show that 61% of respondents are represented by men and 39% by women. However, respondents over 40 years of age (66%) are the most represented, while those under 40 years of age represent 34% in this survey.

The information gathered during this study shows us that, people with primary (42%) and secondary (33%) education are more represented, while illiterate people (24%) and academics (1%), are less represented in this survey. In addition, the most represented socio-professional categories are farmers (65%), while pickers (15%), traders (8%), craftsmen and housekeepers (5%) each, and civil servants (2%) are less represented in this study.

These results reveal that the Ngbaka (59%) and Mbati (20%) ethnic groups are predominantly represented, while the Aka pygmies (12%) and Bofi (9%) are poorly represented in this survey.

IV.1. FLORISTIC DIVERSITY OF PFNL IN THE MUNICIPALITY OF PISSA

This study allowed us to inventory a total of 102 NTFP species belonging to 78 genera and 42 botanical families. It appears from this study that of the total NTFPs inventoried in the commune of PISSA, 50% are used in food, 26% and 24% are used respectively as service plants (other uses) and in the traditional Pharmacopoeia.

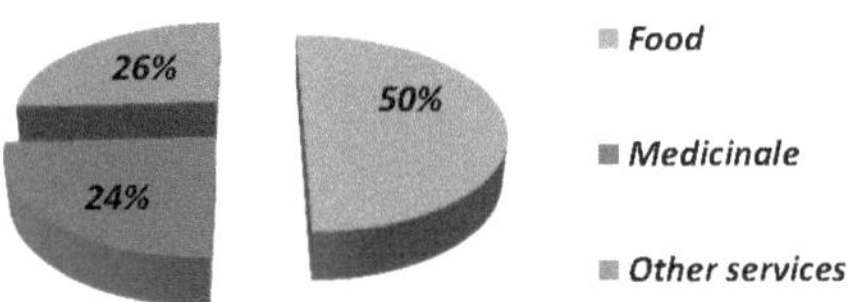

Figure 3: Diagram of NTFP categories used in the commune of PISSA

IV.2 1.Families represented in the municipality of PISSA

This study allowed us to identify 42 families of NTFPs used in the commune of Pissa. The most represented families in this locality are Euphorbiaceae and Sapotaceae (7 species each), Annonaceae and Arecaceae (6 species each), Marantaceae, Moraceae, Sterculiaceae (5 species each) and Poaceae (4 species).

The least represented families in this commune are Irvingiaceae, Cesalpinaceae, Meliaceae and Rubiaceae (3 species each), Anacardiaceae, Asteraceae, Gnetaceae, Clusiaceae, Mimosaceae, Myrtaceae, Ulmaceae, Verbenaceae and Zingiberaceae (2 species each) and the remains (annexes).

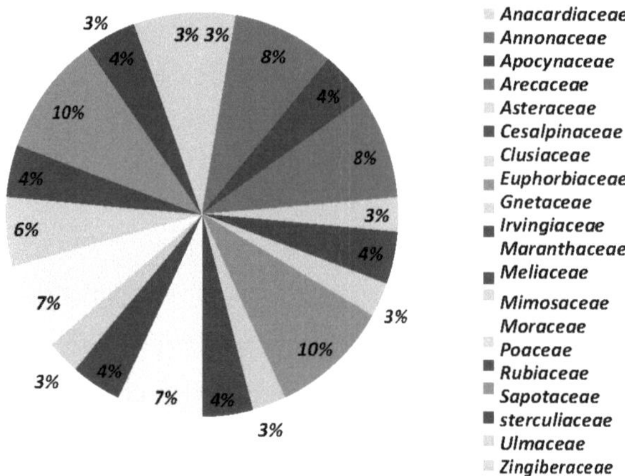

Figure 4: Diversity of NTFP Families in the Municipality of Pissa

IV.2.2. Gender cited in the municipality of PISSA

A total of 102 NTFP species belonging to 78 genera have been recorded in this locality. The most represented genera are : *Cola, Irvingia Hypselodelphys, Gambeya* (each 3 species) and the genera *Gnetum, Garcinia, Morinda Afromomun* have 2 species each, the remains are less represented (fig. 5).

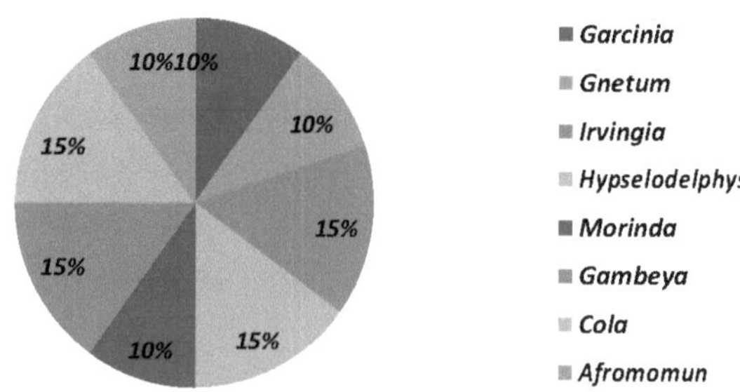

Figure 5: Diagram of the most cited genders in the commune of Pissa

IV.2.3. Species cited in the municipality of PISSA

A total of 102 plant species have been recorded in the municipality of Pissa, the most cited species are : *Gnetum bulchhozianum* (79 citations), *Marantochloa sp* (79 citations), *Oxytenanthera abyssinica* (73 citations), *Alchornea cordifolia* (72 citations), *Dorstenia scaphigera* (74 citations), *Copaifera mildbraedeii* (70 citations), *Amphalocarpum sp* (68 citations), *Afrotyrax lepidophyllus* (67 citations), *Syzygium aromaticum* (66 citations), *Londophia lanceolata* (64 citations), *Borassus sp* (57 citations), *Hilleria latifolia* (56 citations), *Talinum triangulare* (42 citations), *Carica papaya* (21 citations), *Laccosperma secundiflorum* (47 citations) *Drypetes gossweileri* (30 citations), *Musa sp* (25 citations), *Piper guineense*(25 citations), *Gambeya gigantea* (22 citations), *Hypselodelphys poggeana* (21 citations), *Thomandersia laurifolia* (27 citations). This result includes both food, medicinal and other NTFPs (fig. 6).

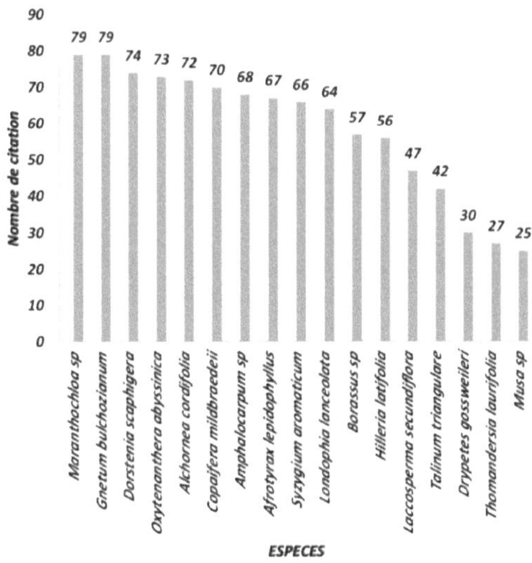

Figure 6: Diagram of the most cited species in the commune of Pissa

IV.2 4.Representativeness of households surveyed on NTFPs

Figure 7 illustrates the results obtained. The best fit of this curve is a logarithmic function of equation y = 40.64ln(x)-9.291 where y is the number of plants and x is the number of respondents. The high correlation coefficient (R^2 = 0.912) illustrates the good fit. Examination of this figure shows that increasing the number of households would do little to actually expand the list of plants used in the commune. Saturation is reached with 60 households.

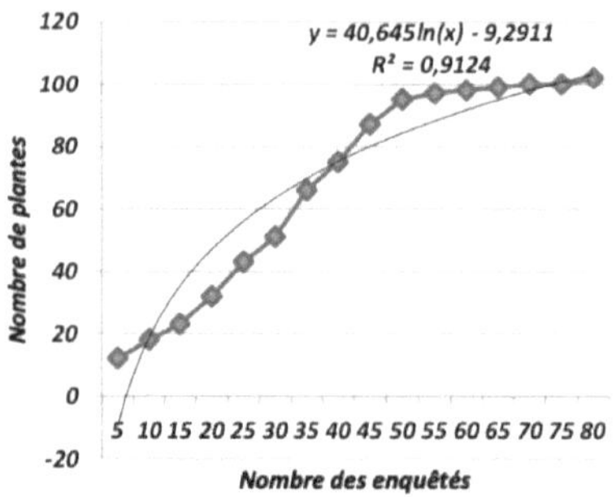

Figure 7: Evolution of the number of plants indicated as a function of the number of households surveyed in the commune of Pissa

IV.2. CATEGORIES OF NFPS IN THE MUNICIPALITY OF PISSA IV.3.1.NFPS FOOD USE

IV3.1.1.Floristic diversity of food NTFPs a.Food NTFP families

The floristic analysis of NTFPs conducted identified 28 NTFP food families. The Sapotaceae family (7 species) and Sterculiaceae (5 species) are the most represented, while the Euphorbiaceae, Irvingiaceae, Annonaceae, Moraceae, Zingiberaceae family (3 species each), Arecaceae, Myrtaceae, Poaceae (2 species each), are the least cited, while the other remaining families have only one species (Fig. 8).

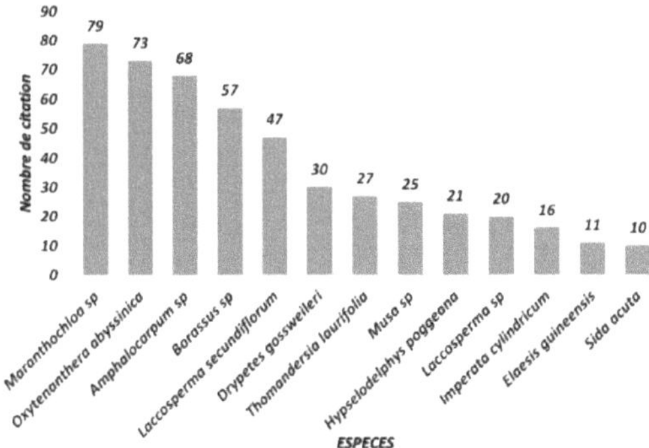

Figure 8: Diagram of NTFP species cited in other services

c. Plant organs and types of use used in other services

We have identified three (3) plant organs and seven (7) types of uses in the category of other services. The most used plant organs are stems (57.14%) and leaves (39.28%), the least cited are whole plants (3.57%) (Fig. 26a).

In addition, the most cited types of uses are agricultural tools (32%) and packaging (23%), while roofs (15%), traditional brooms (12%), basketry (9%), dugout canoes (6%) and mats (3%) are the least cited types of uses (Fig. 26b).

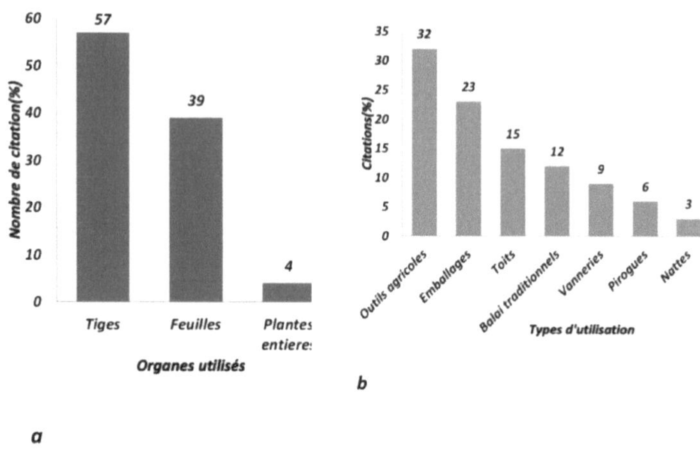

Figure 9: a) Diagram of Plant Organs b) Types of NTFP use in other services

d. Representativeness of the number of households surveyed of plants used in other services

A total of 79 households provided information on the use of plants used in other services in the locality. Figure 27 illustrates the results obtained, the best fit of this curve is a logarithmic function of equation y=7.332ln(x)+8.631, where y is the number of plants and x is the number of respondents. The high correlation coefficient (R^2 = 0.952) illustrates the good fit. Examination of this figure shows that increasing the number of households would do little to expand the list of plants used in other services. Saturation is reached with 60 households. This result illustrates the popularity of the use of these plants used in other services.

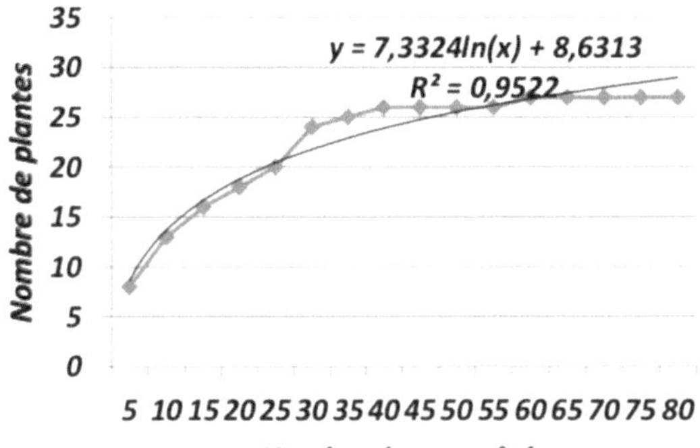

Figure 10: Diagram of the representativeness of the number of households surveyed on NTFPs for service use

IV.3.3.1. MANAGEMENT OF NLP USED IN OTHER SERVICES

a. Harvesting methods and niches used in other services

We identified four (4) modes of harvesting NTFPs used in other services, the most cited modes of harvesting are felling (52%), picking (41%), while harvesting by uprooting (7%) is less cited and no picking (0%) of NTFPs used in other services (Fig 28a).

The majority of NTFPs used in other services are harvested in the forests (59%), and behind the huts (31%) while very little is harvested in the fields (10%), (fig 28b).

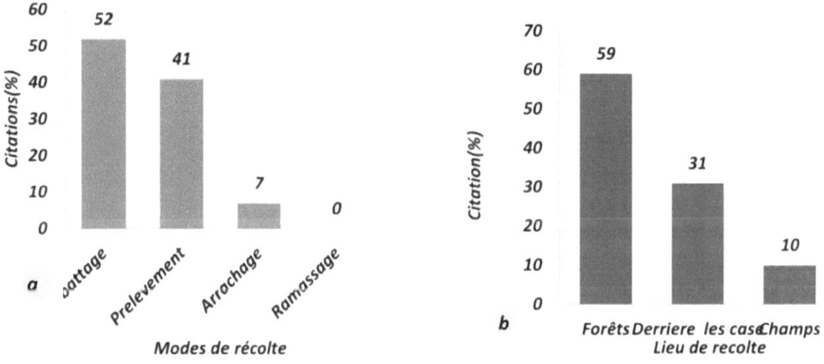

Figure 11: a) Harvesting Pattern Diagram b) NTFP harvest niche used in other services

b.Distance and Condition of Resources at the Place of Harvest of Other Services

The survey shows that the distance to the harvest site of NTFPs used in other services is mostly more than 5 km (68%) from the village, whereas harvesting at a distance of less than 1 km (32%) is less cited by the commune's population (Fig. 29a). Four (4) resource states were assessed by the population in this category during our survey: according to the population, the abundant resources represent 43%, and 48% of the resources are scarce while the less abundant (9%) and 0% are non-existent (Fig. 29b).

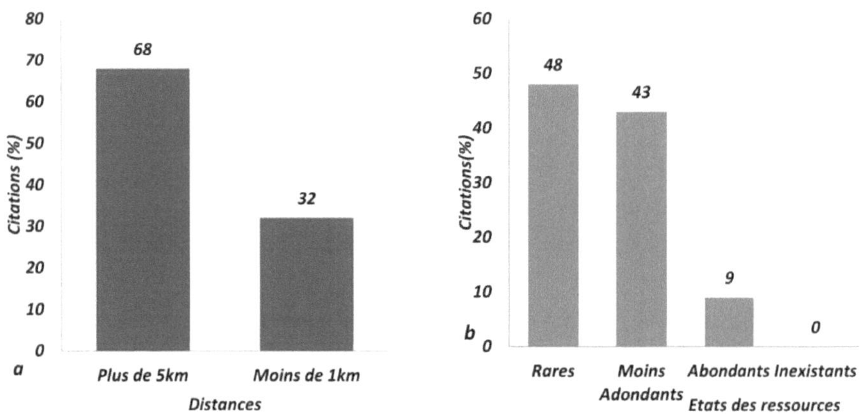

***Figure 12**: a) Diagram of distance to harvest site b) Resource status at harvest site of NTFPs used in other services*

c.Regeneration and Conservation Methods in Other Services

The survey showed that 81% of the plants used in the other services regenerate naturally, while 15% and 4% regenerate by seed and vegetative propagation, respectively (Fig. 30a). However, 77% of the plants used in the services are not conserved, and 23% of the plants are conserved by the traditional method, while no customary law or myth is cited by the surveyed population for the conservation of these resources (Fig. 30b).

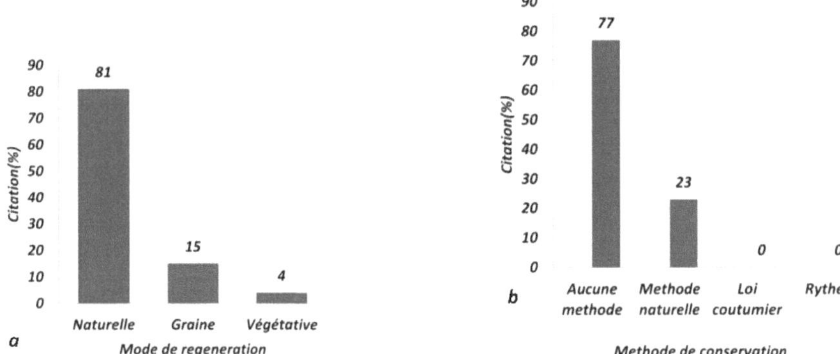

Figure 13: a) Diagram of regeneration patterns b) Conservation methods for NTFPs used in other services

IV.3.4.PFNL FOR MULTIPLE USE

In a global framework, nine (9) NTFPs belong to both the food and traditional pharmacopoeia categories: *Dacryodes edulis, Copaifera mildbraedeii, Hilleria latifolia, Alchornea cordifolia, Irvingia excelsa, Persea americana, Khaya anthotheca, Cymbopogon citratus, Garcinia punctata ;*

Five (5) NTFPs belong to the food and other uses categories: *Elaesis guineensis, Borassus sp, Musa sp, Coffea sp, Amphalocarpum sp, Londophia lanceolata ;*

Two (2) NTFPs are in the categories of traditional pharmacopoeia and other uses *Imperata cylindrica, Gmelina arborea*). Only one (1) NTFP belongs to 3 categories of uses *Carica papaya* (fig 31).

CHAPTER V DISCUSSION AND CONCLUSION

V.1.DISCUSSION

The questionnaire administered to the populations of the commune of PISSA showed in general that the local populations use Non-Timber Forest Products as a source of subsistence. The examination of the results obtained shows the importance that the populations of the said locality give to NTFPs to express their current problems (food, health and other services). Data processing revealed that in this study, men represent 61% and 39% are represented by women, however 66% of the surveyed populations are over 40 years old while those under 40 years old represent 34%. This low representation of the female gender and a non-negligible representation of people over the age of 40 years explains that, wild NTFP species are found much more in the forests far from the villages surveyed and women harvest species located near the village, while men can travel greater distances in the forest and knowledge of plant uses and their properties are generally acquired through long experience accumulated and transmitted from one generation to another. The transmission of this knowledge is currently in danger because it is not always assured (Anyinam 1995). People with primary (42%) and secondary (33%) education have a significant percentage of plant use, while illiterate people (24%) and academics (1%) use plants very little. This result is similar to the results of some ethnobotanical work carried out at the national level, which showed that nearly 40% of village chiefs claim to be able to read and write (Wanyombo, 2010; Ngassé, 2010). In the study area, the vast majority of plant users are farmers (65%), while gatherers (15%), traders (8%), craftsmen and householders (5%) each, and civil servants (2%) are the least represented in this study. In addition, the Ngbaka (59%) and Mbati (20%) ethnic groups are predominantly represented, while the Aka pygmies (12%) and Bofi (9%) are poorly represented. This confirms the results of the surveys conducted by Wanyombo, 2010 and Mbabale, 2011, where the Ngbaka represent more than a third of household members in CAR.

This study allowed us to inventory a total of 102 plant species belonging to 78 genera and 42 botanical families in this municipality. As such, 50% of the NTFPs inventoried are used in food. This figure is close to the study conducted by the FAO in 2003, which estimates that in CAR, 65% of NTFPs are used in food.

Moreover, the comparison of the list of plants identified in this locality and their use to the data in the literature shows that almost all the species identified are also found in other Central African countries: these are *Gnetum bulchhozianum, Irvingia gabonensis, Khaya grandifolia ,cola sp, Elaeis guineesis, Afromomum sanguieum, Marantochloa sp* etc..) (Mbolo, 1990; Debroux & Dathier, 1993; Debroux & al. 1998).

In the category of NTFPs for food use, it appears that the most used plant organs are fruits (40%) and leaves (29%,) the least cited are seeds 11%. In addition, these food NTFPs are used as: snacks (45%), condiments (20%) and vegetables (14%). These results are similar to the results of work carried out in CAR (Bonanée, 2002) and Cameroon (Betti, 2001), in the home gardens of the populations of the Dja Biosphere Reserve, fruits, seeds and leaves are mainly exposed in the markets for their food use.

For Tchatat & Ndoye, the method used to harvest forest fruits is largely influencée by tree pruning, fruiting of shrubs or young productive trees, especially if intended for self-consumption, causes little damage within the forest structure (Tchatat & Ndoye, 2006). However, excessive fruit harvesting can compromise the survival of species and the poor regeneration of some natural resources providing food NTFPs, due to inappropriate and excessive harvesting methods.

In the category of NTFPs for medicinal use, it appears from this investigation that the most used plant organs are leaves (50%), bark (28%) and roots (14%). However, the most common methods of preparation of the products are decoction and maceration and the drugs are administered orally.

The intense use of leaves and bark in traditional pharmacopoeia has already been reported by Betti (2007) in surveys on the sale of medicinal NTFPs in Cameroonian markets. This can be explained by the fact that leaves are the site of photosynthesis and sometimes storage of secondary metabolites responsible for the biological properties of the plant (Bigendako & Lejoly, 1990, Lumbu & al., *2005*; Mangambu & al. 2008; Kumar & Lalramnghinglova, 2011).), but also by the fact that they are quick and easy to harvest (Bitsindou, 1996), hence their important use as medicine.

In the category for other uses, the most used plant organs are stems (57.14%), leaves (39.28%) and the most cited types of uses are : Agricultural implements (32%), Packaging (23%), Roofs (15%), Traditional brooms (12%).This result is similar to the results obtained elsewhere. The intensive exploitation of the stem can have an impact on the plant and its population.

As far as the management of these resources is concerned, the most used harvesting methods are standing harvesting, i.e. 92% 73% respectively for medicinal and food NTFPs. These harvests through standing harvesting have no impact on the survival or regeneration of the species if the frequency of harvesting is reduced (FAO, 2006 b). On the other hand, harvesting in the NTFP category used in other services is by felling (52%), this method of harvesting reduces the productive potential of the forest in the short term and can even affect species richness if it becomes intensive for certain species categories (FAO, 2006 b). The impact of this exploitation on the structure and composition of the forest is closely linked to this intensity but also to the plant organ removed (Tchatat & Ndoye, 2006). Poor methods of harvesting NTFPs is at the root of their disappearance and are the same causes of disappearance encountered in all localities where NTFPs are found. (Cunningham 1996).

In addition the niches of NTFP harvesting are forests, 75% for food NTFPs and 59% in traditional pharmacopoeia and other services. This affirms the results of the work of (Tsagué, 1995 & Tchatat, 1999), that CAR opens almost all of its various permanent forests to local residents, the other countries of the region seriously limit the rights of use, the severity of the impact of the mode of exploitation varies, depending on the harvesting technique of the organ harvested. In fact, NTFPs in open access areas are the responsibility of the village community and are, therefore, much more vulnerable than those where access is controlled, and farmers take care to preserve interesting NTFP species when clearing the forest (Tchatat, 2002). For Clark & Sunderland in 2004, NTFPs are collected from quite diverse habitats They come from wild and/or cultivated and domesticated species and most of them come from forest, non-forest areas such as grasslands and agricultural fields, and cocoa plantations (Sunderland & al., *2000).*

I want morebooks!

Buy your books fast and straightforward online - at one of world's fastest growing online book stores! Environmentally sound due to Print-on-Demand technologies.

Buy your books online at
www.morebooks.shop

Kaufen Sie Ihre Bücher schnell und unkompliziert online – auf einer der am schnellsten wachsenden Buchhandelsplattformen weltweit! Dank Print-On-Demand umwelt- und ressourcenschonend produziert.

Bücher schneller online kaufen
www.morebooks.shop

KS OmniScriptum Publishing
Brivibas gatve 197
LV-1039 Riga, Latvia
Telefax: +371 686 204 55

info@omniscriptum.com
www.omniscriptum.com

Printed by Books on Demand GmbH, Norderstedt / Germany